Rodolfo Barragan Ramirez
Andres Gonzalez Hernandez
Marcos Alfredo Azuara Hernandez

Analysis of the mechanical properties of concrete with oyster shells

Rodolfo Barragan Ramirez
Andres Gonzalez Hernandez
Marcos Alfredo Azuara Hernandez

Analysis of the mechanical properties of concrete with oyster shells

Management with the use of natural resources in concrete in compression and flexural tests

ScienciaScripts

Imprint

Cover image: www.ingimage.com

This book is a translation from the original published under ISBN 978-613-9-40680-7.

Publisher:
Sciencia Scripts
is a trademark of
Dodo Books Indian Ocean Ltd. and OmniScriptum S.R.L publishing group

120 High Road, East Finchley, London, N2 9ED, United Kingdom
Str. Armeneasca 28/1, office 1, Chisinau MD-2012, Republic of Moldova, Europe
Managing Directors: Ieva Konstantinova, Victoria Ursu
info@omniscriptum.com

Printed at: see last page
ISBN: 978-620-8-55518-4

Contents

Chapter 1

INTRODUCTION.

1.1 BACKGROUND

The history of concrete is the history of man's search for a space to live in with the greatest possible comfort, safety and security. Ever since man overcame the age of huts, he has applied his greatest efforts to delimit his living space, first satisfying his housing needs and then erecting constructions with specific requirements. Temples, palaces, museums are the result of the effort that constitutes the basis for the progress of mankind. The Egyptian people already used a mortar - a mixture of sand and cement - to bind stone blocks and slabs together when choosing their astonishing constructions.

The history of the use of chemical admixtures in concrete dates back to the last century, after Joseph Aspdin patented in England on 21 October 1824, a product he called "Portland Cement".

The first addition of calcium chloride as an additive to concrete was registered in 1873 and patented in 1885. At the same time as accelerators, the first additives used were water-repellent. Likewise, at the beginning of the century, the incorporation of sodium silicate and various soaps was tested to improve waterproofing. At that time, fine powders began to be added to colour the concrete. Fluates or fluosilicates were used from 1905 onwards as surface hardeners. The retarding action of sugar had also already been observed.

1.1.1 BASIC COMPONENTS OF HYDRAULIC CONCRETE.

Aggregates are generally divided into two groups: fine and coarse. Fine aggregates consist of natural or manufactured sands with particle sizes ranging from 15 mm to approximately 10 mm; coarse aggregates are those whose particles are retained in the No. 16 mesh and can be up to 152 mm. The maximum aggregate size commonly used is 19 mm or 25 mm.

The binder is composed of Portland cement, water and entrapped or intentionally included air. Typically, the paste constitutes: 25% to 40% of the total volume of concrete.

As aggregates constitute approximately 60% to 75% of the total volume of concrete, their selection is important not because of the volume percentage, but because they are the building blocks.

The characteristics of the composition of its particles must have an adequate resistance, as well as resistance to weathering conditions, because if they were to contain impurities, they could cause deterioration of the concrete. In order to have an efficient use of the binder (cement and water, air), it is necessary to have a continuous granulometry in particle sizes. The quality of the concrete depends largely on the binder.

In a properly made concrete, every aggregate particle is covered with aggregate in its full dimension, as are all spaces between aggregate particles.

1.1.2 COMPRESSIVE STRENGTH.

The less water used, the better the quality of the concrete, as long as it can be properly consolidated. Lower amounts of mixing water result in stiffer and more difficult to manipulate mixes; but with vibration, even these mixes can be easily manipulated.

For a given quality of concrete, the stiffest mixes are the most economical. Therefore, consolidation of concrete by vibration allows for an improvement in concrete quality and economy.

1.1.2 BENDING STRENGTH.

Specifications and investigations of apparent low strengths should take into account the high variability of flexural strength test results. A short drying period can produce a sharp drop in flexural strength. Flexure can be used for design purposes, but should be tested with care as they are extremely sensitive.

1.2 PROBLEM DEFINITION.

Nowadays, the concepts of ecology and environment are becoming more and more important worldwide, and this directly affects the construction industry because the type of activities that involve this industry can have harmful and even irreversible consequences on the environment. In recent years, scientific publications, patents and websites of some countries or companies have been carrying out studies and exposing the advances in the production, characterisation and applications of oyster shell components. It has been found that materials that are the product of calcareous species can be considered as natural hybrid materials, as they are made up of organic and inorganic compounds. An example of a natural hybrid material is bone, whose hardness and stiffness has not been found in any synthetic material.

The advantage that these porous materials offer is that, by removing the organic components, the size and shape of the pores can be precisely controlled, making them fire resistant, impermeable in construction and for environmental uses, such as absorbing pollutant gases.

One way to make good use of crustacean waste is to add it to the binder as it is a good quality substitute for calcium carbonate and contributes to increasing the mechanical strength of the concrete.

1.3 OBJECTIVE OF THE RESEARCH.

1.3.1 GENERAL OBJECTIVE.

To evaluate the mechanical properties of concrete in flexural and compression tests with the addition of oyster shell (Crassostrea).

1.3.2 SPECIFIC OBJECTIVES.

Determine the index properties of aggregates (gravel, sand and oyster shell) for use in the manufacture of hydraulic concrete and analyse their mechanical properties in relation to conventional concrete.

1.4 RESEARCH QUESTIONS.

1) Which of the two concretes will give higher compressive strength?
2) Increase workability?
3) adjust other concrete properties?
4) Will it increase or decrease its slack?
5) Will its unit mass increase or decrease?

1.5 HYPOTHESIS.

Do the properties of concrete improve with the addition of oyster shell (Crassostrea)? Are the flexural and compressive test performances different compared to a conventional concrete mix?

1.6 JUSTIFICATION.

At present, the "Crassostrea" oyster shells, commercially exploited in the Moralillo Municipality of Panuco, Veracruz, represent a pollution problem. After consuming the edible part of the oyster, people discard the shells in the streets or in public places and even in landfills, causing a bad appearance, poor hygiene and the spread of pests.

Therefore, this research aims to provide a solution to this pollution problem and recycle this waste to obtain a material for environmental purposes for construction.

1.7 DELIMITATIONS.

1.7.1 GEOGRAPHICAL DELIMITATIONS.

The natural material of oyster shell (Crassostrea) was obtained from the Laguna de la Costa which is located within the locality of Moralillo, in the Municipality of Pánuco (in the State of Veracruz de Ignacio de la Llave). It is located exactly 38.2 km (W) from the geographic centre of the municipality of Pánuco. And it is located 0.69 km (to the W) from the centre of the town of Moralillo.

Figure 1.1 Location of oyster (Crassostrea) shell.

Samples of oyster shells were taken to the laboratory to be crushed and sieved, with particle sizes ranging from 75 micrometres (mesh № 200) to 4.75 millimetres (mesh № 4), and may contain smaller fines, within the proportions established in standard N-CMT-2-02-002/02.

1.7.2 TECHNICAL DELIMITATIONS.

Furthermore, its capture and production is the main activity for the inhabitants of the state's coastal lagoons ; the problem persists when it is not known what to do with the inedible part of this marine product. This problem affects restaurateurs, the civilian population and the authorities, since tons of them are generated and only a quarter of them are used for the reproduction of the adductor mollusc. This is not exclusive to the state, but at a national level, other coastal states such as Tamaulipas, Veracruz and Campeche are also affected by this same problem.

Disposing of the shells in these places generates considerable inconvenience, mainly because of their insolubility in water and their resistance to biodegradation. One of the problems that most worries the communities is the growth of the dengue-carrying mosquito inside the oyster shells, as they have an irregular and asymmetrical shape, whose outer side is rough and dark, contrasting with the inside, which has a smooth, concave surface that allows water to be retained in it.

This project was carried out in approximately 6 months, starting in November 2018, and finishing at the end of May 2019. Although the research could be more in-depth and extensive, finalising details in each of the components of the project, it will be up to future students interested in expanding this research further to do so.

6 cylinders of 15 x 30 and 9 beams of 15x15x60 cm of concrete were manufactured with Cemex Monterrey CPC-30R Portland cement marketed in the Zona Conurbada del Sur de Tamaulipas, with virgin aggregates from Cacalilao Veracruz River Sand and crushed pebble gravel of % " to fines from Banco el Abra in the State of SLP with the addition of 3% by weight of oyster shell residue.

In order to obtain their resistance, compression and bending tests were carried out with a Forney universal machine, model LT-1150, which has different capacity ranges.

The properties of the concrete we evaluated were: in the fresh state slump test (SLUMP), temperature, unit mass, and in the hardened state its compressive and flexural strength at 28 days.

The following equipment and materials were used for manufacturing and testing:

- Cement.
- Water.
- Arena.
- Equipment:
- Steel or wrought iron moulds (before use the moulds should be coated with mineral oil or a non-reactive form release agent).
- Iron rod.
- Rubber mallet.
- Wheelbarrow.
- Steel construction shovels.

They were manufactured:

- 6 concrete cylinders 15*30cm
- 9 beams 15*15*60 cm

Chapter 2

ANALYSIS OF FUNDAMENTALS.

2.1 THEORETICAL FRAMEWORK.

2.1.1 BACKGROUND TO CEMENT.

Of all the hydraulic binders, Portland cement and its derivatives are the most widely used in construction because they are basically made up of mixtures of limestone, clay and gypsum, which are very abundant minerals in nature, their price is relatively low compared to other materials, and their properties are very suitable for the goals they are intended to achieve.

The hydraulic binders also include blast-furnace cements, pozzolanic cements and mixed cements, all of which have a very wide field of use in concrete for certain media, as well as aluminous cements "calcium aluminate cements", which are used in special cases.

Cements are used to produce mortars and concretes when mixed with water and aggregates, natural or artificial, to obtain prefabricated or "in situ" construction elements.

5,000 years ago, in northern Chile, the first stone structures appeared, joined together by a hydraulic conglomerate made from the calcination of seaweed. These structures formed the walls of the huts used by the Indians.

The Egyptians used gypsum and lime mortar in their monumental constructions.

In Troy and Mycenae, history says that stones joined by clay were used to build walls, but, actually, concrete made with a minimum of technique appears in vaults built one hundred years before J.C.

The Romans took an important step forward when they discovered a cement made by mixing volcanic ashes with quicklime. At Puteoli, known today as Puzzuoli, there was a deposit of these ashes, hence this cement was called "pozzolana cement".

Agrippa built the Pantheon in Rome with concrete in 27 BC, which was destroyed by fire and later rebuilt by Hadrian in 120 AD. Since then, it has stood the test of time without damage until 609, when it was transformed into the church of Santa Maria dei Martiri (Saint Mary of the Martyrs). Its 44-metre-long dome is made of concrete and has no openings other than a skylight at the top.

2.1.2 BACKGROUND TO THE AGGREGATE.

Throughout the history of mankind, man has needed to learn how to use nature's own resources. Thus, the history of aggregate processing can be traced back to the activity that has taken place in the earth's interior throughout the geological eras, leading to changes in the formation and transformation of the rocks that are used today in the production of concrete, asphalt mixes and pavement structures.

In the last 100 years, the need for building materials has led man to study natural laws and understand them through careful observation of rocks from their natural state to their means of use.

Aggregate is a natural and/or artificial solid, stony material of apparently inert, stable form, of variable composition, pH and size.

Aggregates have the following characteristics according to their shape and texture, as these are very influential in the water demand of the mix and in the adhesion of the aggregate to the cement paste:

- Absorption (NTC 176): Measures surface porosity or saturable porosity. The more porous it is, the less mechanical resistance it has, so the less absorption it has, the better its compaction and quality.
- Number of particles passing the 0.075 mm sieve (NTC 78) or Pass 200: Particles smaller than 0.075 mm interfere with the adhesion between the aggregate and the cement paste or asphalt, causing them to lose their binding capacity.
- Density (NTC 176): Indicates the ratio between the mass and volume of the rock.
- Abrasive wear (NTC 93, 98): Indicates the resistance of the material when rubbed against another material of known resistance or by rubbing the stone particles against each other.
- Shape of aggregates (NTC 385): Classification of aggregates according to their shape: elongated, flat, round and fractured. The greater adherence to the cement paste and asphalt is due to the greater fracturing, producing greater interlocking of particles and therefore greater resistance.
- Granulometry (NTC 77 and 174): It is the composition in percentage of the various sizes of aggregates in a sample. This composition is made from the largest to the smallest size, by a figure that represents, in weight, the partial percentage of each size that passed or was retained in the different sieves that are compulsorily used for

such measurement.

- Moisture (NTC 1776): Estimates the degree of surface or free moisture mainly in fine aggregate.
- Unit mass (NTC 92): This works out the ratio between the mass of the material that fits in a given container and the volume of that container. It is expressed in kg/m3.
- Alkali-silica reactivity (NTC 175): Generation of the alkali-silicate gel resulting from the reaction of the active silica components of the aggregate and the alkalis in the cement. This gel is expansive and when absorbing water it tends to increase in volume causing internal pressures with the corresponding cracking and rupture of the cement paste.
- Sanitation (NTC 126): Determines the soundness of aggregates when subjected to the action of weathering or aggressive environmental conditions.
- Harmful substances (NTC 127, 589): Coarse aggregates must be clean without particles of organic or inorganic nature, as they decrease the strength and economy of concretes and asphalts.

2.1.3 EMERGENCE OF PORTLAND CEMENT.

In the mid-17th century, the English Reverend James Parker accidentally created a cement by burning some limestone. This discovery was christened Roman cement because it was then thought to have been used in Roman times, and it began to be used on a number of building sites in the UK.

Joseph Aspdin and James Parker patented on 21 October 1824 the first Portland cement made from clayey limestone and coal: calcined at high temperature. The name Portland comes from its greyish colour, very similar to the stone of the Isle of Portland in the English Channel.

Later Isaac Johnson improved this production process by increasing the calcination temperature, obtaining in 1845 the prototype of modern cement made from a mixture of limestone and clay calcined at high temperatures until the formation of clinker. This is why Johnson is known today as the modern father of Portland cement.

Towards the end of the 19th century, a number of developments from the period of propelling the use of Portland cement for a wide range of applications. Important in this case was the development of industrialisation, the introduction of rotary kilns for calcination and the tube mill.

Similarly, the cement industry experienced rapid growth at the beginning of the 20th century, due to the experiments of the French chemists Vicat, Le Chatelier and the German Michaelis, who succeeded in producing a cement of homogeneous quality.

The above summarises the main conditions for the production of portland cement in large quantities for the construction industry at the turn of the century and beyond.

Today, despite all the technical improvements that have been introduced, Portland cement is still, in essence, very similar to what it was first proposed, although its impact and presentation have improved significantly.

Table 2.1 Chronology of the cement industry in Mexico

YEAR	EVENT
1906	The first Mexican cement plant was founded in Hidalgo, N.L., which later became what today is the "CEMEX GROUP".
1928	Holcim Apasco is founded in the municipality of Apasco, State of Mexico.
1931	Cementos Hidalgo and Cementos Portland Monterrey merge to form Cementos Mexicanos, now CEMEX.
1942	Creation of the Cement Regulatory Commission.
1942	Establishment of the cement industry office.
1948	The national cement chamber (CANACEM) is created with the participation of all companies incorporated as public limited companies.
1959	The Mexican Institute of Cement and Concrete is founded, which takes on the functions of dissemination of cement and concrete, as well as participation in the elaboration of cement and concrete quality standards.
1973	The first yearbook is published that includes relevant information about on cement production and consumption in Mexico, as well as data on the industry.
1992	An agreement is signed with the IMSS to differentiate the cement industry from the gypsum and lime industry in order to pay more realistic contributions.
1994	The National Occupational Health and Safety Award is signed

	jointly with the IMSS. The SCT was able to get the SCT to restart the use of concrete in road pavements, thus ending a myth that lasted 70 years. CANACEM participates as a founding member of the National Organisation for Standardisation and Certification of Construction and Building (ONNCCE).
2008	Holcim Apasco starts construction of a new cement plant in Hermosillo, Sonora.

Source: Estudio y aplicación Normativa en la fabricación del cemento, Isaí Montalván; Leny Suarez; Edith Téllez, Instituto Politécnico Nacional, Mexico DF 2010.

2.1.4 TYPES OF CEMENT.

The classification of cements can be made according to different criteria, in order to In this project we will focus on their classification according to the type of cement. Portland.

There are five types of Portland cement:

- Type I: Portland cement intended for concrete works in general, when the use of another type is not specified (buildings, industrial structures, housing complexes). It releases more hydration heat than other types of cement.
- Type II: of moderate sulphate resistance, Portland cement intended for concrete works in general and works exposed to moderate sulphate action or where moderate heat of hydration is required, when specified (bridges, concrete pipes).
- Type III: High initial strength such as when the concrete structure needs to be loaded as soon as possible or when formwork needs to be stripped within a few days of pouring.
- Type IV: Low heat of hydration is required where no expansion should occur during setting.
- Type V: Used where high resistance to concentrated sulphate action is required (canals, sewers, harbour works).

2.1.5 MECHANICAL STRENGTH.

Regarding strength, Espino et al (1997) say that it is the most important property of the structural material, as it is the one that defines the force that a structural element will be able to withstand before it fails, and that this is known as stress.

The mechanical strength of cement is the main characteristic that the user evaluates and appreciates. When cement hydrates with water, it constitutes the matrix that ensures the strength of the skeleton of aggregates that make up mortars and concretes. The intrinsic strength of cement is an increasing function of the calcium silicate content in the clinker and the fineness of grind, as basic parameters.

The increase in strength over time depends on the ratio between the C3S that generates the initial strength and the C2S that contributes later. In hardened pastes, irrespective of the cement's own strength, the strength is due to the volume of hydration products formed in the space defined by the cement and the mixing water. This factor is to some extent expressed in the classic water/cement ratio.

The strength of the asphalt lies in the paste-aggregate bonding conditions and/or in the intrinsic strength of the paste.

It should be remembered that the strength of the aggregates far exceeds the strength of the paste and that which the concrete elements normally have to assume.

The strength of concrete differs according to the type of stress imposed on it: for example, in compression it resists ten times more than in attraction.

With compressive strength being the highest, concrete has a natural vocation to meet this working regime; being reinforced with steel bars that take up the tensile stresses or undergoing a state of pre-stressing that compensates for the tensile stress.

On the other hand, compressive strength is a general index of quality as it correlates with the modulus of elasticity and is an efficient indicator of durability. It is clear that compression test results are of interest as concrete is judged by its compressive strength. Especially, when the testing standards have incorporated plastic monteros with strengths comparable to those obtained in concrete.

2.1.6 COMPRESSIVE STRENGTH TESTS

Cement compressive strength, the most common test applied to concrete, is an extreme test measure used in determining the appropriate cement type and application of different grades of Portland cements.

Its compressive strength is a fundamental measure of safe design. The compressive strength test measures the ability of concrete to resist compressive forces.

From the combination of cement, water, sand and gravel the concrete is created, specialists test the samples in concrete cylinders under controlled conditions in order to determine the compressive strength.

The hydration of concrete occurs at different rates, so that it reaches different levels of strength in different periods of time. Concrete hydration takes place at different rates, so that concrete reaches different strength levels in different time spans.

Tests can be performed at one day, three days, seven days, 28 days, or 90 days. After that each sample is hardened in the given time.

Specialists place under a compression load, in a hydraulic machine, until it reaches the breaking point.

Compressive strength is measured by taking cylindrical specimens of concrete in the compression testing machine, while compressive strength is calculated from the breaking load divided by the area of the section resisting the load and reported in mega Pascals (MPa) in SI units. Compressive strength requirements can range from 17 MPa for residential concrete to 28 MPa and higher for commercial structures.

In the next step the results are analysed according to the specified compressive strength requirements according to the standards. The compressive strength of concrete mixtures can be designed in such a way that they have a wide range of mechanical and durability properties that meet the structural design requirements.

The results of the compressive strength tests are primarily used to determine that the supplied concrete mix meets the specified f'c strength requirements of the project.

The results of strength tests from moulded cylinders can be used for concrete acceptance quality control purposes or to estimate the strength of concrete in structures for scheduling construction operations, such as formwork removal or to assess the suitability of curing and protection provided to the structure.

Cylinders subjected to acceptance and quality control testing are made and cured following the procedures described in standard cured specimens. A test result is the average of at least two standard or conventionally cured strength tests made from the same concrete sample and tested at the same age.

To perform this test, according to the construction materials practice guide of the Arturo Narro Siller Faculty of Engineering (FIANS) at the Autonomous University of Tamaulipas (UAT), a press with a load capacity compatible with the strength of the elements, applied at a speed of 25kg/cm per second, is required.

The sample to be tested must be in a state of humidity in equilibrium with the environment, recommending a storage period of no more than 4 days in the laboratory, with air circulation around the specimens.

Prior to the test it is necessary to have the total area and the net area of each specimen to be tested, the net area of each sample is that which is included in the chamfers, a layer of sulphur is applied on the upper and lower layers of the block to uniform the load to which it will be subjected.

This sulphur layer must have a higher compressive strength than the specimen, otherwise the sulphur would fail first and the results obtained would be erroneous.

The load is applied without impact and uniformly until the limit at which the load cannot be sustained. The maximum reading is recorded in the register.

2.1.6 BENDING STRENGTH TESTS.

Flexural strength is a measure of the tensile strength of concrete (concrete). It is a measure of the resistance to moment failure of an unreinforced concrete beam or slab. It is measured by applying loads to concrete beams 6 x 6 inches (150 x 150 mm) in cross section and spanning at least three times the thickness.

It is expressed as the Modulus of Rupture (MoR) in pounds per square inch (MPa). The modulus of rupture is approximately 10% to 20% of the compressive strength, depending on the type, dimensions and volume of the coarse aggregate used.

Never allow the beam surfaces to dry out at any time. Soak in water saturated with lime for at least 20 hours before testing.

2.2 CONCEPTUAL FRAMEWORK.

2.2.1 CONCEPT AND DEFINITION OF CEMENT.

Cement is a hydraulic binder, i.e. a finely ground inorganic material which, when mixed with water, forms a paste which sets and hardens through reactions and hydration processes and which, once hardened, retains its strength and stability even under water.

Properly dosed and mixed with water and aggregates, it should produce a concrete or mortar that retains workability for a sufficient time, achieves pre-specified strength levels and exhibits long-term volume stability.

Hydraulic hardening of cement is mainly due to the hydration of calcium silicates, although other chemical compounds, e.g. aluminates, may also be involved in the hardening process. The sum of the proportions of reactive calcium oxide (CaO) and reactive silicon dioxide (SiO2) shall be at least 50 % by mass, where the proportions are determined in accordance with European Standard EN 196-2.

Cements are composed of different materials (components) which, properly dosed through a controlled production process, give the cement the physical, chemical and resistance qualities suitable for the desired use.

There are, from the point of view of standardised composition, two types of components:

Main component: Inorganic material, specially selected, used in a proportion greater than 5% by mass of the sum of all main and minority components .

Minority component: Any major component, used in a proportion of less than 5% by mass of the sum of all major and minority components.

2.2.2 CEMENT MANUFACTURE: INDUSTRIAL ACTIVITIES IN CEMENT MANUFACTURE.

2.2.2.1 FIRST STAGE: RAW MATERIALS.

The cement manufacturing process starts with the studies and mining evaluation of raw materials (limestone, clays, sand, iron ore and gypsum) necessary to achieve the desired metal oxide composition for the production of clinker. Cement clinker is composed of the following oxides (data in %)

Table 2.2 Components of clinker

CLINKER COMPONENTS	PERCENTAGE (%)
Calcium carbonate (CaCO3)	70-75
Silicon oxide (Si02)	2-3
Aluminium Oxide (AL203)	20-25
Iron Oxide (FE03)	0-1

Source: http://www.canacem.org.mx/procesos_de_produccion.htm _3/10/2014

The right proportion of the different oxides is obtained by dosing the starting minerals.

- Limestone and marl for CaO supply.
- Clay and shales for the rest of the oxides.

As a second step, the geological studies are completed, the exploitation is planned and the process begins: drilling, burning, removal, classification, loading and transport of raw materials.

The essential raw materials, limestone, marl and clay, which are extracted from quarries, must provide the essential elements in the cement manufacturing process: calcium, silicon, aluminium and iron.

Very often other secondary raw materials, either natural (bauxite, iron ore) or by-products and residues from other processes (thermal power plant ashes, steel slag, foundry sands) must be used to provide these elements. Limestones can be hard enough to require the use of explosives and then crushing, or soft enough to be exploitable without the use of explosives.

Once the large masses of stone have been fragmented, the resulting material is transported to the plant in trucks or belts. The natural raw materials are subjected to an initial crushing, either in the quarry or on arrival at the cement factory where they are unloaded for storage.

Figure 2.1 Shredding

The crushing of the rock is carried out in two stages, initially it is processed in a primary crusher, of the cone type, which reduces it from a maximum size of 1.5m to 25cm. The material is deposited in a stockpile. Then, after verifying its chemical composition, it goes to secondary crushing, reducing its size to approximately 2 mm.

The crushed material is transported to the actual plant by conveyor belts and deposited in a raw material stockpile. In some cases, a pre-homogenisation process is carried out. Pre-homogenisation through proper stacking designs and extraction of materials in storage reduces the variability of materials.

This material is transported and stored at a site from which the raw mill is fed. Two more silos with corrective materials (iron ores and high corrective limestone) are stored there. They are dosed depending on their characteristics; and by means of scales the material to the flour (or raw) mill. The studies of the composition of the materials in the different quarry areas and the analyses carried out in the factory make it possible to dose the mixture of raw materials to obtain the desired composition.

2.2.2.2.2 SECOND STAGE: GRINDING AND COOKING.

This stage involves the grinding of raw materials (raw grinding), by ball mills, roller presses or high compression forces, which produce a very fine material. In this process, the selection of materials is carried out, according to the design of the mixture, in order to optimise the raw material that will enter the kiln, considering the cement with the best characteristics.

Grinding is used to reduce the particle size of the materials so that the chemical reactions during firing in the kiln can be carried out properly. The mill grinds and pulverises the materials to an average size of 0.05 mm.

The ground material must be homogenised to ensure the effectiveness of the clinkerisation process through consistent quality. This process is carried out in homogenising silos. The resulting material, which is a very fine powder, must have a constant chemical composition.

The kiln must receive a chemically homogeneous feed, which is achieved by controlling the correct dosage of the materials that make up the feed to the raw mill. If the materials used are variable in quality, they must be pre-homogenised beforehand.

After the mill, the raw meal undergoes a final homogenisation process, which ensures a homogeneous mixture with the required chemical composition.

In addition to chemical homogeneity, the fineness and particle size curve of the raw meal is essential. This is achieved by adjusting the separator that classifies the product leaving the mill, reintroducing the insufficiently ground phase (closed circuit).

Figure 2.2 Grinding and Firing Process.

The second stage also consists of firing the raw meal in rotary kilns until it reaches a material temperature of around 1450 °C to be cooled abruptly and obtain an intermediate product called clinker.

2.2.2.3 THIRD STAGE: CLINKER MANUFACTURE.

Clinker is defined as the product obtained by incipient fusion of clayey and calcareous materials containing calcium oxide, silicon, aluminium and iron in suitably calculated quantities.

Clinker is an intermediate product in the cement making process. A source of lime such as limestones, a source of silica and alumina such as clays and a source of iron oxide are properly mixed, finely ground and calcined in a kiln at approximately 1500°C, resulting in the so-called Portland cement clinker.

The raw meal is introduced by means of pneumatic transport systems and duly dosed into a multi-stage gas suspension heat exchanger, at the base of which a modern pre-calcination system is installed for the mixture before it enters the rotary kiln where the remaining physical and chemical reactions that lead to the formation of clinker take place.

The heat exchange takes place by means of thermal transfers by intimate contact between the material and the hot gases obtained from the furnace, at temperatures of 950 to 1100°C.

The kiln is the fundamental element for the manufacture of cement. It consists of a cylindrical steel tube with lengths of 40 to 60 m and diameters of 3 to 6 m, which is lined on the inside with refractory materials. The kiln for cement production produces temperatures of 1500 to 1600°C since the clinkerisation reactions are around 1450°C. The clinker that exits the kiln at a temperature of 1200°C then undergoes a rapid cooling process using grate coolers. It is then transported by metal conveyors to a storage area.

Depending on how the material is processed before it enters the clinker kiln, four types of manufacturing processes can be distinguished: dry, semi-dry, wet and semi-wet.

The technology applied depends mainly on the origin of the raw materials. The point of limestone and clay and the water content (from 3% for hard limestones to 20% for some marls) are the decisive factors.

Dry process:

The dry process is the most economical in terms of energy consumption and is the most common. The raw material is fed into the kiln in a dry, powdery form. The kiln system comprises a cyclone tower for heat exchange, which preheats the material in contact with the gases coming from the kiln.

The limestone decarbonation (calcination) process can be almost completed before the material enters the kiln if a combustion chamber is installed to which fuel is added (precalciner).

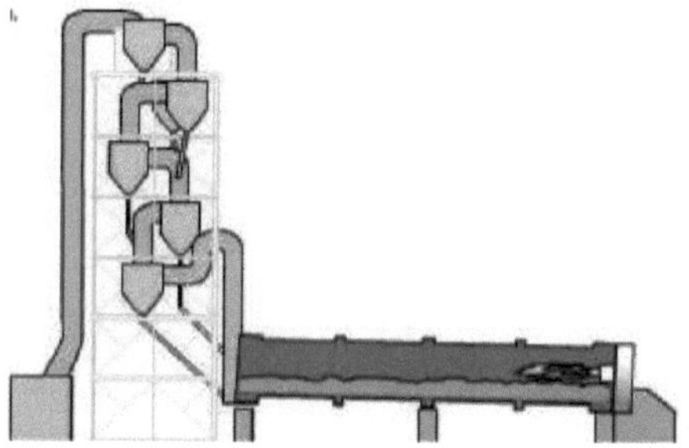

Figure 2.3 Calcination

Wet processing:

This process is normally used for raw materials with a high moisture content. The feed material is prepared by co-milling the same water, resulting in a slurry with a water content of 30-40% which is fed into the high end of the clinker kiln.

If the clay is quite wet and has the property of dissolving in water, it must be subjected to the action of mixers to form the slurry; this is done in the washing mill, which is a circular shaft with radial stirring arms with rakes which break up the agglomerates of solid matter.

Semi-wet and semi-dry process.

The feed material is obtained by adding or removing water, respectively, to the material obtained in the raw milling process. Pellets or granules with 15-20% moisture content are obtained, which are deposited on mobile grates through which hot gases from the furnace are circulated.

When the material reaches the kiln inlet, the water has evaporated and firing has begun. In all cases, the material processed in the rotary kiln reaches a temperature of around 1450°C. It is cooled abruptly as it leaves the kiln in planetary or grate coolers to obtain clinker.

Cement grinding:

The cement manufacturing process ends with the joint grinding of clinker, gypsum and other materials called "additions". The materials that can be used, which are standardised as additions, are among others:

- Blast furnace slag.
- Natural pozzolans.
- Fly ash.
- Limestone.

Depending on the composition, strength and other factors, cement is classified into different types and classes. Cement grinding is carried out on mechanical equipment where the mixture of materials is subjected to the impact of metallic bodies or high compressive forces. The following equipment is used for this purpose:

- Roller press.
- Vertical roller mills.
- Ball mills.
- Horizontal roller mills.

It is important to know the different stages in the cement manufacturing process, which is considered the industrial activity and is schematized by identifying the different activities.

2.2.3 MAIN USES AND APPLICATIONS OF CEMENT.

Nowadays, when we mention the word cement, it indicates any type of adhesive in civil engineering construction, as well as a substance that can be expanded to join sand and crushed rock and some type of aggregate to form a solid mass. Its properties include greater strength and mechanical resistance than the aggregates that gave rise to it.

A cement can be a unified chemical compound, but in most cases it is a mixture, although from a chemical point of view it is generally a mixture of calcium silicates and aluminates, obtained by firing limestone, clays and sands, from which different types of cement are obtained.

In table 2.3, they are distinguished by their composition, strength properties, durability, by their intended uses and uses.

Table 2.3 Uses and applications of cement

AREA	MAIN USE
Improving housing	Improvement and renovation of housing, foundations, pavements, patios, etc.
Agricultural	Resists environmental aggressiveness
Industrial	Reinforced with plastic or metal fibres to reduce cracking
Marine/Hydraulic	With low permeability characteristics and high resistance under contact with water
Special applications	Joint sealant, on sheet metal or asbestos-cement roofs

Fuente:http://tesis.ipn.mx/bitstream/handle/123456789/7781/I2.1198.pdf?sequ ence=119/12/2014

2.3 REGULATORY FRAMEWORK.

2.3.1 CEMENT TYPES, SPECIFICATIONS AND LABORATORY TEST METHODS.

MEXICAN STANDARD NMX-C-414-ONNCCE-2004: Construction industry - hydraulic cements - specifications and test methods.

Objective and scope: This standard establishes the specifications and test methods applicable to the various types of hydraulic cements of national or foreign manufacture intended for consumers in Mexico.

This standard is supplemented by the following existing Mexican standards:

NMX-C-059-0NNCCE: Construction industry - Determination of setting time of hydraulic cementitious materials.

NMX-C-061-ONNCCE: Construction industry - cements - determination of compressive strength of hydraulic materials.

The SCT standard N-CMT-02-002/02 CMT characteristics of materials - materials for structures - materials for hydraulic concrete - quality of stone aggregates for hydraulic concrete.

It contains the quality characteristics of aggregates used in the manufacture of hydraulic concrete with the exception of lightweight aggregates that were used for the production of fireproof concrete, as well as fillers and concrete elements whose design is based on load testing and not on conventional procedures.

The SCT standard N-CMT-2-02-005 CMT characteristics of materials - materials for structures - materials for hydraulic concrete - quality of hydraulic concrete. This standard contains the quality characteristics of hydraulic concrete to be used in the construction of structures.

SCT manual M-MMP-2-02-001/00 MMP Materials Sampling and Testing Methods - Materials for Structures - Materials for Hydraulic Concrete - Portland Cement Concrete Sampling.

2.3.2 TYPES OF CEMENTS.

2.3.2.1 CLASSIFICATION ACCORDING TO THEIR SPECIAL CHARACTERISTICS.

The characteristics of the cements are: resistance to sulphates, low alkali-aggregate reactivity, low heat of hydration and white colour.

The respective cements should have an additional designation according to their special characteristics.

Sulphate Resistant (RS) Cements: Those cements which, due to their behaviour, comply with the requirement of limited expansion according to the established test method.

Low Alkali-Aggregate Reactivity Cement (LAC): Low Alkali-Aggregate Reactivity Cement (LAC): Low Alkali-Aggregate Reactivity Cement (LAC) meets the requirement of limited expansion in the alkali-aggregate reaction, according to the established test method.

Low heat of hydration (LHHH) cement: Cement that develops a heat of hydration equal to or lower than that specified in the standard.

White cements (B): These are all cements whose whiteness index must be equal to or higher with reference to the established standard.

2.3.4 CLASSIFICATION OF CEMENT BY STRENGTH CLASSES. The different cements are also classified by compressive strength into five classes according to table 2.4 below.

Table 2.4 Different types of cement

Type	Title	Resistant class	Special features
CPO	Ordinary portland cement	20	Resistance to sulphates.
CPP	Pozzolanic Portland Cement	30	Low aggregate alkali reactivity.
CPEG	High homogeneous granulated slag cement	30 R	Under heat of hydration.
CPC	Composite portland cement	40	White.
CPS	Portland cement with silica fume	40 R	
CEG	Portland cement with granulated slag		

Source:http://www.revistacyt.com.mx/images/problemas/2009/pdf/JULIO.pd1 /10/2014

Standardised designation: Cements should be identified by type and strength class as specified in the table above. If the cement has a specified 3-day strength, the letter R (Rapid Strength) shall be added. If a cement has any of the special characteristics listed in Table II.5, its designation is completed in accordance with the nomenclature indicated in that table, if it has two or more special characteristics.

2.3.5 PHYSICAL SPECIFICATIONS OF CEMENTS.

The physical specifications of cement such as compressive strength according to NMX-C-061-ONNCCE, setting time according to NMX-C-059-ONNCCE and volume stability according to NMX-C-062-ONNCCE are summarised in table 2.5.

Table 2.5 Physical specifications of cement

Physical specifications							
Resistant class	Compressive strength (N/mm2)			Setting time (min)		Volume stability in autoclave (%)	
	3 days minim 0	28 days minim 0	maximum	Minimum initial	Initial maximum	Maximum expansion	Maximum contraction
20		20	40	45	600	0.8	0.20
30		30	50	45	600	0.8	0.20
30R	20	30	50	45	600	0.8	0.20
40		40		45	600	0.8	0.20
40R	30	40		45	600	0.8	0.20

Source:http://www.revistacyt.com.mx/images/problemas/2009/pdf/JULIO.pd1 /10/2014

In cases where the properties of the cement can be improved by exceeding the sulphate (SO3) limits, it is permissible to exceed these limits as long as it does not cause expansion greater than 0.020% after 14 days of immersion in water according to NMX-C-131-ONNCCE and NMX-C-185-ONNCCE. The following table 2.6 shows the chemical specifications.

Table 2.6 Cement chemical specifications

Chemical specifications		
Properties	Types of cements	Specification (% mass)
Loss on ignition	CPO, CEG	Max 5% Max 5% Max
Insoluble residue	CPO, CEG	Max 4% Max 4% Max 4% Max 4% Max 4% Max 4% Max 4% Max 4% Max 4% Max 4% Max 4% Max 4%
Sulphate (SO2)	All	Max 4% Max 4% Max 4% Max 4% Max 4% Max 4% Max 4% Max 4% Max 4% Max 4% Max 4% Max 4%

Source:http://www.revistacyt.com.mx/images/problemas/2009/pdf/JULIO.pd1 /10/2014

When a cement is required to have a special characteristic, it must comply with the specifications indicated in table 2.7. It is verified by methods to determine the chemical characteristics based on NMX-C-418-ONNCCE, NMX-C-180-ONNCCE and NMX-C-151-ONNCCE.

Table 2.7 Specification of cement with special characteristics

Specifications of cements with special characteristics.							
Nomenclatura	Special features	Expansion due to sulphate attack (Max %)	Expansion by alkali-aggregate reaction (Max %)		Heat of hydration (Max) kj/kg (Kcal/kg)		White (min %)
		1 year	14 days	56 days	7 days	28 days	
RS	Sulphate resistant	0.10	——		-		-
BRA	Low reactivity Alkali added	-	0.020	0.060	-		-
BCH	Low hydration heat	-	-	-	250	290	-
B	White	-	- -		-		70

Source:http://www.revistacyt.com.mx/images/problemas/2009/pdf/JULIO.pd1 /10/2014

2.3.6 TECHNICAL INFORMATION - CEMENT REGULATIONS.

- Ordinary Portland Cement - NMX-C-414-ONNCCE-199:

Ordinary portland cement is excellent for general constructions: footings, columns, beams, castles, slabs, walls, slabs, floors, pavements, pavements, municipal furniture such as benches, fountains, tables, ideal for making prefabricated products such as planks, paving stones, blocks, lamp posts, sinks, etc.

- Composite portland cement NMX-C-414-ONNCCE-199:

It has excellent durability in sewer prefabrications and provides concretes with higher chemical resistance and less heat release.

This cement is compatible with all conventional building materials such as sand, gravel, quarry, marble, etc. As well as additive pigments, as long as they are used with the care and dosage recommended by their manufacturers.

- Pozzolanic portland cement NMX-C-414-ONNCCE-199:

Ideal for the construction of footings, floors, columns, castles, dallas, walls, slabs, pavements, pavements, municipal furniture. Especially for construction on saline soils, best for construction sites exposed to chemically aggressive environments.

High durability in prefabricated sewerage products such as manhole covers, storm drains, manholes and drainage pipes.

- White ordinary portland cement NMX-C-414-ONNCCE-199 :

Excellent for ornamental or architectural works such as façades, monuments, tombstones, railings, staircases, etc. Great performance in the production of terrazzo mosaics, balustrades, washbasins, toilets, rural, tiroles, stickers, joints, etc.

The facades of wall coverings, saves repainting costs. This product can be easily pigmented to obtain the desired colour and can be mixed with conventional building materials, as long as they are free of impurities. Due to its high compressive strength it has the same structural uses as grey cement.

- Sulphate resistant ordinary portland cement NMX-C-414- ONNCCE-199:

Sulphate-resistant ordinary portland cement provides higher chemical resistance for concrete in contact with water or aggressive soils such as seawater, soils with high sulphate or salt content. Recommended for the construction of dams, municipal drains, and all types of underground works.

- Portland cement for masonry (Mortar) NMX-C-021-ONNCCE- 2004:

Specially designed for masonry work, jointing or bonding of blocks, partitions, bricks, stone, masonry, plastering, screeding, rendering, rendering and patching, screeds, templates and pavements, not to be used in the construction of structural elements.

Chapter 3

RESEARCH METHOD.

3.1 RESEARCH APPROACH.

The methodological approach is quantitative, as data collection and analysis are used to answer the research questions and test previously established hypotheses, it is a deductive process, each stage leads logically to the next, it serves to prove, explain or predict a certain fact.

3.1.1 TYPE OF RESEARCH.

The type of research is experimental, because it is possible to evaluate in what way or for what reason something in particular happens. This type of research is provoked, which allows the variables to be modified in intensity, and the causes and consequences of the results can be evaluated. The experiment is a voluntary manipulation of variables and the results are observed in a controlled environment.

3.1.2 RESEARCH METHOD.

Material from the selected bench was sampled and the laboratory tests required in the standards were carried out to determine if the material is within the parameters set out in the standard, then the concrete specimens were made and tested, the results obtained were written up and a conclusion was reached.

3.1.4.1 TESTING OF COARSE AGGREGATES

Granulometry test.

The test consists of passing the sample through these meshes and determining the percentage of material retained on each mesh.

Specific gravity test.

The determination of the dry volumetric mass of the material in its loose state consists of obtaining the ratio between the mass of the solids in the material and the total volume of the material, once the mass of the sample has been corrected for water content.

From the sample of the material, obtained according to the Manual M-MMP-1-01, Sampling of Materials for Earthworks, of the Secretariat of Communications and Transport (SCT), the necessary quantity to fill the 10 L container is dried, disaggregated and separated, according to the Manual M-MMP-1-03, Drying, Disaggregation and Quartetting of Samples, of the SCT.

1) The material is homogenised by mixing, and then, using the sheet ladle and the scantling as a reference, the sheet container is filled as shown in Figure #, for which the material is dropped from a height of 20 cm, avoiding its rearrangement by undue movement. Subsequently, the material is levelled using the 30 cm ruler.

2) The mass of the container with the material is obtained as shown in figure # and recorded as W in g, to the nearest 5 g.

3) Finally, the water content of the material is determined according to manual M-MMP-1-04, Water content, which is recorded as w.

Calculations.

The dry volumetric mass of the loose material is calculated and reported as a test result using the following expression:

$$\gamma d_s = \frac{100\, W_m}{V\,(100 + w)} = \frac{\gamma_m}{100 + w}(100)$$

3.1.4.2 TESTING OF FINE AGGREGATES.

Granulometry test:

The test consists of passing the sample through these meshes and determining the percentage of material retained on each mesh.

Table 3.1 Particle-size limits for fine aggregate

Mesh		Percentage retained accumulated
Opening mm	Designation	
9.5	3/8"	0
4.75	№4	0-5
2.36	№8	0-20
1.18	№16	15-50
0.6	№30	40-75
0.3	№50	70-90
0.15	N°100	90-98

Source: http://www.imcyc.com/revistacyt/pdfs/problemas27.pdf

1) A representative sample of the sun-dried, disaggregated and cracked soil is obtained, weighed and the weight is recorded in the corresponding register.

2) The material is passed through the different meshes, which go from the largest to

the smallest opening, as shown in the register for this test.

3) The material retained on each screen is weighed and recorded in the retained weight column.

4) All of the above is done up to the No. 4 mesh and with the material that passes this mesh, a representative portion of soil is obtained, for which the material must be passed as many times as necessary through the sample match, until a sample of between 500 and 1000 grams is obtained.

5) The previous sample is put to dry completely (until it does not fog up the watch glass), it is cooled and a sample of 200.0 grams is weighed, which is poured into an aluminium cup and water is poured into it until it is full; with this, the soil is washed. If the soil under study has an appreciable amount of lumps, it is left in saturation for 24 hours.

6) Soil washing consists of shaking the soil using the wire rod with a rounded tip, making figures in the shape of "eights" for 15 seconds.

7) The liquid is poured into the 200 mesh, in order to remove the fines (which is the material that passes through the 200 mesh), then more water is poured into the beaker and stirred as described above.

8) When a lot of material (sand) accumulates in the mesh, it is reintegrated into the vessel, emptying water over the back of the mesh, always taking care not to lose material; this is done every 5 times that water with fines is emptied into mesh No.200. This operation is repeated as many times as necessary so that the water comes out clean or almost clean.

9) The soil is dried in the oven or kiln, allowed to cool and then passed through the following screens, which are No. 10 to No. 200. For more efficient vibrating, it is recommended that the whole set of screens is passed through the screen vibrator.

10) The material retained on each mesh is weighed.

11) Calculations are made for: % partially retained, % accumulated retained, % passing; the granulometric curve is drawn.

12) The following are calculated: the % of gravel, sand and fines, as well as the Coefficients of Uniformity (Cu) and Curvature (Cc).

3.1.4.3 FRESH CONCRETE TESTING

Slump test:

The slump test is done to ensure that a concrete mix is workable. The measured sample must be within an established range, or tolerance, of the intended slump.

The hydraulic concrete sample, obtained in accordance with Manual M-MMP-2-02-055, Hydraulic Concrete Sampling, shall require no preparation other than remixing for homogenisation.

In this test, reliable slump values are obtained in the range of 2 to 20 cm. The entire operation from the beginning of the filling until the mould is lifted shall be carried out without interruption, in a time not exceeding 2.5 min and according to the following procedure:

1) The inside of the mould is moistened and placed on the previously moistened metal plate.

2) By resting his feet on the mould's stirrups, the operator holds the mould firmly in place and proceeds with the filling operation.

3) The mould is filled in three layers of approximately equal thickness, each layer being compacted by 25 penetrations of the rod evenly distributed over its section.

4) Once the compaction of the last layer has been completed, the concrete is screeded by rolling the rod over the upper edge of the cone. The outer surface of the base plate is cleaned and immediately the mould is carefully lifted up in the direction of vertical, without lateral or torsional movements. The operation of completely lifting the mould shall be done in 5 plus minus 2 sec.

5) Immediately afterwards, the slump of the concrete is determined from the original level of the upper base of the mould by calculating this difference in height at the slumped centre of the upper surface of the specimen.

Compression test:

Commercial or laboratory prepared sulphur mortars used to cover the parallel faces of moulded specimens shall be verified to have a minimum strength of 35,34 MPa (350 kg/cm^{12}) within a maximum time of 2 hours .

1) The surfaces of the upper and lower press plates and the ends of the test specimens are cleaned; the specimen to be tested is placed on the lower plate, aligning its axis

carefully with respect to the centre of the spherically seated load plate as shown in Figure 3.1 while the upper plate is lowered towards the specimen until a smooth and even contact is achieved.

2) The load is applied with a uniform and continuous velocity without producing impact or loss of load. The velocity shall be within the range of 137 to 343 kPas/s (approximately 84 to 210 kg/cm^3/min).

3) The loads are applied until the maximum permissible load is reached, making the corresponding records.

The strength of concrete specimens is determined at the age of 14 days in the case of rapid strength concrete and 28 days when using normal strength concrete, with the tolerances given in the following table.

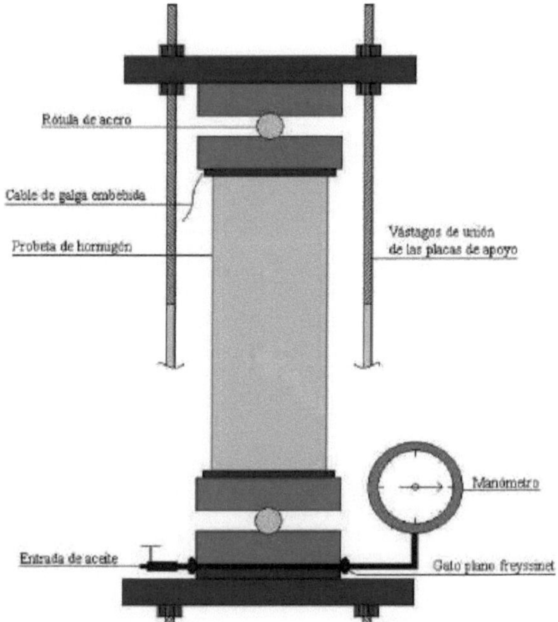

Figure 3.1 Universal machine diagram

Table 3.2 Tolerance in hours

Age of test (days)	Tolerance (hours)
14	± 12
28	± 24

Source: Manual M-MMP-2-02-055

Calculations

As a result of this test, the compressive strength supported by the specimen is calculated and reported, using the following expression:

$$R = \frac{10P}{A}$$

Where:

R = Simple compressive strength, (MPa)

P = Maximum load, (kN)

A = Average cross-sectional area of the specimen, (cm^2)

3.1.6 METHOD OF DATA ANALYSIS.

For the data analysis of the present study, 12 concrete cylinders AND 9 concrete beams made under the same mix design, with the only variation being the addition of oyster shell (Crassostrea), were subjected to compression and flexural tests.

Chapter 4

ANALYSIS OF RESULTS.

This chapter shows the results obtained from the laboratory tests.

For research purposes, two dosages were carried out, based on the ratio:

1:2 (1 cement: 2 Sand)

water/cement ratio Ra/c= 0.67

The dosages were as follows:

Mix № 1 cement, sand, gravel and water.

Mix № 2 cement, sand, gravel, water and 3% oyster shell.

The materials were homogeneously mixed and controlled in weight, to try to ensure equal conditions in the specimens sampled in cylindrical moulds of = 15 cm h=30 cm and rectangular moulds of 15 x 15 x 60 cm; the moisture content of the mixtures in general, temperature, volumetric masses in the fresh state, percentage of absorption, slump, compressive stresses at pre-established ages were determined.

4.1 MATERIALS USED IN THE MANUFACTURE OF HYDRAULIC CONCRETE

As part of the research, prior to the design itself, the quality characteristics of the aggregates used in the manufacture of hydraulic concrete were determined.

According to the Ministry of Communications and Transport, aggregates are selected natural stone materials; materials subject to disintegration, screening, crushing or washing treatments, or materials produced by expansion, calcination or excipient fusion, which are mixed with Portland cement and water to form hydraulic concrete. Aggregates for hydraulic concrete are classified as: fine aggregate and coarse aggregate.

4.2 FINE MATERIAL IN ITS NATURAL STATE.

The fine aggregate used in the production of hydraulic concrete is selected natural sand or obtained by crushing and screening, with particle sizes between 75 micrometres (mesh № 200) and 4.75 millimetres (mesh № 4), and may contain fines of smaller size, within the proportions established in the standard N-CMT-2-02-002/02.

4.1.1 SAND GRANULOMETRY.

The purpose of particle size analysis of sand is to determine the quantities in which particles of certain sizes are present in the material.

The particle size distribution is carried out by using square aperture meshes of the following sizes: 3/8", Numbers 4, 8, 16, 20, 20, 30, 40, 40, 50, 60, 60, 80, 100 and 200 respectively, table 4.1 shows the percentage of retained and passed.

Table 4.1 Percentage of retained fines

Mesh	Partially retained		Retained cumulative % Retained cumulative % Retained cumulative % Retained cumulative % Retained cumulative % Retained	Passing %	Specifications % passing
	Grs.	%			
3/4	7.71	1.30	1.30	98.70	
4	23.95	4.05	5.36	94.64	95 a 100
8	20.71	3.50	8.86	91.14	80 a 100
16	22.52	3.81	12.67	87.33	50 a 85
30	79.69	13.48	26.15	73.85	25 a 60
50	272.04	46.02	72.17	27.83	10 a 30
100	109.55	18.53	90.70	9.30	2 a 10
p-100					
sum	591.15	100.00	100.00		

The test results are plotted together with the limits specifying the acceptable percentages for each size to verify whether the size distribution is appropriate.

The most suitable particle size for the fine aggregate depends on the type of work, the richness of the mix (cement content) and the maximum size of the coarse aggregate.

Graph № 1 shows the results of the granulometry test carried out on the material used in the manufacture of the hydraulic concrete, material from the Cacalilao bench, Veracruz.

Grain-size curve of bench material

Cacalilao

Malla No.	ARENA
	% que pasa la malla
3/8"	100.00
No. 4	99.00
8	98.00
16	92.00
30	69.00
50	16.00
100	2.00
200	10.00

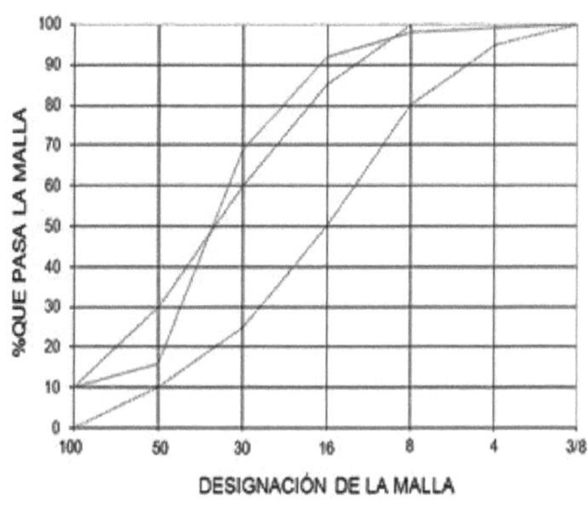

As can be seen in the graph above, the portion of the material corresponding to meshes 30 and 16 is outside the acceptance parameters for use in the manufacture of hydraulic concrete.

4.1.2 CALCULATION OF THE MODULUS OF FINENESS

The granulometric analysis of the sand is complemented by calculating the fineness modulus which is equal to one hundredth of the sum of the accumulated retained percentages in each of the meshes of the standard series. The sand is usually considered to have a fineness modulus suitable for the manufacture of concrete.

MF

Z%Ret.Accum. on meshes (3/4"; #4; #8; #16; #30; #50; #100) - 100

Sands with a fineness modulus of 2.24 are fine sand; and if the modulus is between 2.3 and 3.1, it is medium sand. If the modulus is higher than 3.1, it is a coarse sand.

When they have a fineness modulus lower than acceptable, they are considered too fine and are detrimental to this application, as they tend to require higher cement paste consumption, which adversely affects volumetric changes and the cost of concrete.

At the other extreme, sands with a fineness modulus greater than 3.5 are too coarse and are also judged unsuitable because they tend to produce concrete mixes that are coarse, segregable and prone to bleeding.

The sand tested in the laboratory has a fineness modulus of 2.24, which is at the specification limit, which means that it is a fine sand, this sand can be used for the manufacture of cement, with favourable results.

The loose dry volumetric weight of 1241 kg/m3 was also obtained.

The absorption percentage is in the order of 5.52% with a specific gravity of 2.427.

4.2 PETREOUS MATERIAL IN ITS NATURAL STATE

The results obtained from testing the material from bench #0040 "El Abra" located at kilometre 081+600 of the Cd. Valles - Cd. Victoria road under natural conditions are presented below.

4.2.1 PARTICLE SIZE OF VIRGIN COARSE AGGREGATES

Now, the particle size and maximum aggregate size are important because of their effect on the batching, workability, economy, porosity and shrinkage of concrete.

For the gradation of coarse aggregates a series of meshes are used which are specified in the Mexican Standard N-CMT-2-02-002/02.

Table 4.2 Coarse aggregate particle sizes

Number de Malla	Retained (gr)	Partial % % Partial % Partial % Partial % Partial % Partial % Partial % Partial	Retained cumulative % Retained cumulative % Retained cumulative % Retained cumulative % Retained cumulative % Retained	Pass the mesh %
2"				
1.5				
1	50	0.86	0.86	99.14
3/4	1,170	20.17	21.03	78.97
1/2	2,145	36.97	58.00	42.00

3/8	975	16.80	74.80	25.20
№4	1,035	17.84	92.64	7.36
№8	140	2.41	95.05	4.95
P8				
SUMA	5,802	100		

Coarse aggregate, according to the SCT, can be selected natural gravel or obtained by crushing and screening, air-cooled blast furnace slag or a combination of these materials, with maximum particle sizes generally between 19 mm (3/4") and 75 mm (3"), and may contain rock fragments and sand, within the proportions established in the standard N-CMT-2-02-002/02.

Graph № 2 shows the results obtained from the granulometry test carried out on material from El Abra bench, San Luis Potosí.

Graph № 2 Grain-size curve of material from El Abra Bench

Mesh No.	GRAVA
	% passing mesh
2 1/2"	100.00
2"	100.00
1 1/2"	100.00
1"	100.00
3/4"	85.00
1/2"	45.00
3/8"	22.00
No. 4	2.00

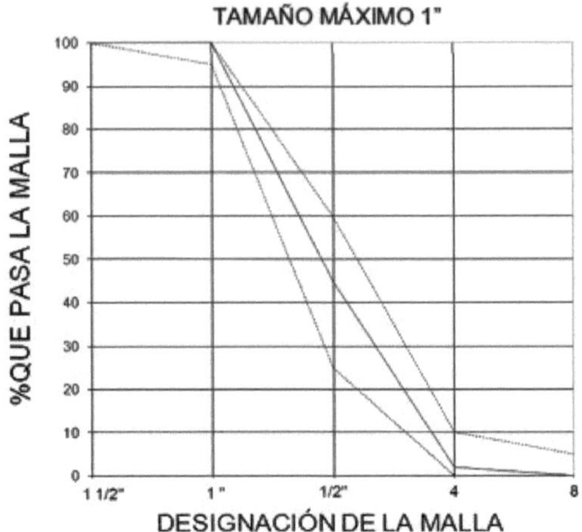

According to the parameters established by the SCT, the material presents an acceptable behaviour to be considered as coarse aggregate in concrete with a maximum size of 1".

Tests were carried out on the resistance to wear by abrasion in the angel machine and accelerated weathering, with the results being: 29.50% and 5.50% respectively.

The dry volumetric weight was also obtained as 1450 kg/m3 dry loose and 1528 kg/m3 dry dry weight.

The absorption percentage is in the order of 0.55 % with a specific gravity of 2.692.

Portland Cement CPC 30R

CEMEX brand CPC 30R portland cement was used to make the hydraulic concrete.

This cement can be used in the construction of all types of simple or reinforced concrete elements or structures. It is compatible with all conventional construction materials, achieving excellent results in the traditional construction of: floors, floors, floors, castles, beams, footings, slabs, columns, etc. CPC 30R is produced under a strict quality control that makes it the cement of excellent application for all types of works, from family projects to the construction of subdivisions, houses, buildings, municipal works, industrialised concrete products, etc. It is a type of cement that is available in the area in any commercial shop dedicated to supplying construction materials.

Water

The water used for the production of concrete is commonly distributed by the municipal drinking water commission of the municipalities of Tampico, Ciudad Madero and Altamira. This water is obtained from a lagoon system known as Chairel, which is formed by the Tamesí-Pánuco rivers, and is free of oils, acids, alkaline substances and organic matter (treated water).

4.3 PREPARATION OF RECYCLED MATERIAL "OYSTER SHELL" CRASSOSTREA ANGULATA

The results obtained from testing recycled oyster shell material (Crassostrea Angulata) obtained from the Laguna de la Costa which is located within the locality of Moralillo, in the Municipality of Pánuco (in the State of Veracruz de Ignacio de la Llave) are presented below.

Samples of oyster shells were taken to the laboratory to be crushed and sieved, with particle sizes ranging from 75 micrometres (mesh № 200) to 4.75 millimetres (mesh № 4), and may contain smaller fines, within the proportions established in standard N-CMT-2-02-002/02.

Dosage of the mixture

Once the characteristics of the aggregates were obtained, a dosage was made to respect the established water-cement ratio of 0.67, as well as the percentage of recycled material.

The dosage used is shown below:

Table № 1 Dosage of mixtures

Dosage	Cement o	Sand Kg.	Grav a	Agu a	Con ch a
1	20.00	47.00	61.00	13.41	-
2	44.00	102.00	134.00	32.00	9.18

Remark:

Volumetric weights were considered:

cement = 1250 Kg/m3;

sand= 1241 Kg/m3

sand absorption= 5.52%,

water= 1000 kg/m3,

Oyster shell = 940 kg/m3 in compact state,

gravel= 1450 Kg/m3

gravel absorption = 0.55%.

The specimen processing date was 21 March 2018 with an ambient temperature of 26.4°C and relative humidity of 80%.

The proportions calculated by any method should always be considered as subject to revision on the basis of experience gained with test mixtures.

Depending on the circumstances, test mixtures can be prepared in a laboratory, or perhaps preferably as a field test mixture.

The mixture was prepared according to the following procedure:

The quantities of each of the materials to be used were weighed and placed in 20-litre containers.

The required water was measured in test tubes.

The necessary equipment for sampling and testing was prepared and greased, such as a slump cone, thermometer, equipment for the determination of
volumetric mass, etc.

It was ensured that the 0.6 m3 capacity stirrer was working properly, at the same time the mixing drum was moistened and the excess water was completely drained off.

Prior to the start of the rotation, the coarse aggregate and some water were added.

The mixer was switched on and the fine aggregate, cement and the rest of the water were added with the mixer running. If this is impractical for a particular mixer or for a particular test. These components (fine aggregate, cement and water) can be added to the mixer stopped and allowing the mixer to rotate a few revolutions, continuing the loading with the coarse aggregate and some water.

After all components were in the mixer, the timer was timed: 3 minutes of mixing, followed by a 3 minute rest period (the open end of the mixer was covered to prevent evaporation during the rest period), followed by a final mixing period of 2 minutes.

The concrete was poured into a pre-moistened pan.

With the concrete in the tray, the slump test was carried out and the cylinder and beam moulds were filled.

Note: Start the slump test within five minutes of obtaining the concrete sample. For the moulding of the samples, 15 minutes are allowed from the manufacture of the concrete. According to NMX-C-156.

It should not be mixed for a longer or longer period than specified, as there will be evaporation of water in the mix, with a consequent decrease in workability and increase in strength. Another side effect is the crushing of the aggregates, especially if they are not hard, the particle size becomes finer and the workability lower.

Table № 2 Mixing data

Dosage ón	Moisture in mixture	Revenirmento	Temperatora	Weight Volumetrics
1	11.81	9.0	25.0	2296
2	12.52	7.0	27.3	2290

Note: The volumetric weight of the mixture was determined by the procedures laid down in NMX-C-138-ONNCCE.

The results obtained from the testing of the conventional concrete samples are presented below.

Table № 3 Results of the Mix № 1

Spéci men №	Age in days	Diámetro cm	Height cm.	Load Kg	Compressive stress
1	7	15.1	30.0	30000	167.50
2	7	15.1	30.0	30500	170.30
3	14	15.1	30.0	32250	180.07
4	14	15.1	30.0	32000	178.67
5	28	15.1	30.0	39500	220.55
6	28	15.1	30.0	41000	228.92

Table № 4 Mixing Results № 1

Spécimen	Age in days	Width (a) cm	Height (h)	Long (1) c	Support spacing (L)	Load (P)	Bending stress

№			c			Kg	Kg/cm^2
7	7	15.0	15.0	60.0	45.0	1500	20.00
8	14	15.0	15.0	60.0	45.0	1750	23.33
9	28	15.0	15.0	60.0	45.0	2250	30.00

The table below shows the formation of cement, sand, gravel, water and 3.0% oyster shell.

Table № 5 Results of the

Mixing № 2

Spéci men №	Age in days	Diámetr 0 cm	Height cm.	Load Kg	Compressi ve stress
10	7	15.1	30.0	22250	124.23
11	7	15.1	30.0	22500	125.63
12	14	15.1	30.0	27800	155.22
13	14	15.1	30.0	27400	152.99
14	28	15.1	30.0	29500	164.71
15	28	15.1	30.0	30500	170.30

Table № 6 Results of the Mix № 2

Espécim at №	Age at days	Width (a) -ein	Height (h) -ein	Larg o(l) cm	Distance between apoyos /T \	Charge (P) Kg	Effort or bending Kg/cm^2
1 6	7	15.0	15.0	60.0	45.0	1450	19.33
1 7	7	15.0	15.0	60.0	45.0	1400	18.67
1 8	14	15.0	15.0	60.0	45.0	1850	24.67
1	14	15.0	15.0	60.0	45.0	1900	25.33

9							
2 0	28	15.0	15.0	60.0	45.0	1950	26.00
2 1	28	15.0	15.0	60.0	45.0	2000	26.67

COMMENTS

1. Specimen processing.

During the process and elaboration of the mixture for specimen formation, 2 dosages were carried out, which were described and assigned as № 1 and 2 in the Introduction of this document.

Normal setting behaviour and loss of plasticity was observed during the conventional mix, while for the oyster shell mix, the setting time was longer than 24 hours.

Representative portions of fresh concrete were obtained, in accordance with the hydraulic concrete sampling methods manual M- MMP-2-02-055, in addition to the sampling work, mould filling, packaging and identification, using a total of 12 cylinders with a height/diameter ratio = 2, 0 = 15 cm and 30 cm high; 6 samples of conventional concrete and 6 with oyster shell, for testing at ages of 7, 14 and 28 days; 9 beams of 15 x 15 x 60 cm for bending tests; 3 of conventional concrete and 6 with the natural resource.

The moisture content of each mix was determined in the fresh state; mix № 1, which represents the conventional concrete, presented a moisture content of 11.81%; the case for mix № 2 was in the order of 12.52%.

The specimens were cured by total immersion in water in a curing tank at room temperature.

Prior to testing, the specimens were head-loaded to ensure that the load application faces were maintained perpendicular to the specimen axis; the simple compressive strength of the cylindrical specimens was determined by load application on a FORNEY model LT-1150 universal machine, in an application range of 75000 kg with minimum appreciation of 250 kg at an application rate of 15000 kg per minute.

2. Stress values obtained on specimens at the age of 7 days.

The compressive stress obtained for the mixture № 1 corresponds to 169 kg/cm2.

Tests were carried out to determine the bending stresses with beams of section 15.0 x 15.0 x 60.0 cm, obtaining values according to the expression PL/bd2, where P is the axial load applied, L the length or span of supports, b the base and d the height; for this mixture a value of 20 kg/cm2 was obtained.

For the mixture № 2 corresponding to 3% oyster shell, a compressive stress of 125 kg/cm2, bending stresses of 19 kg/cm2.

Since the failure occurred within the middle third of the specimen, it was not necessary to use correction factors.

3. Stress values obtained on specimens at the age of 14 days.

The compressive stress obtained at the age of 14 days for the mix № 1, corresponds to 179 Kg/cm2.

The bending stresses obtained by applying the load to the beams gave a value of 23.33 kg/cm2.

For the mixture № 2 corresponding to 3% oyster shell, a compressive stress of 154 Kg/cm2, bending stress of 25 Kg/cm2 is presented.

4. Stress values obtained on specimens at the age of 28 days.

A final test was carried out at the age of 28 days.

1 corresponds to 225 kg/cm2, while the bending stresses were 30.00 kg/cm2.

For the mix № 2, a compressive stress at the age of 28 days of 168 kg/cm2 was obtained, the bending stresses obtained were 26 kg/cm2.

Below is a graph of stress vs. age in days in which the evolution of the concrete samples can be observed.

Graph N° 1. Compressive stress vs. age in days

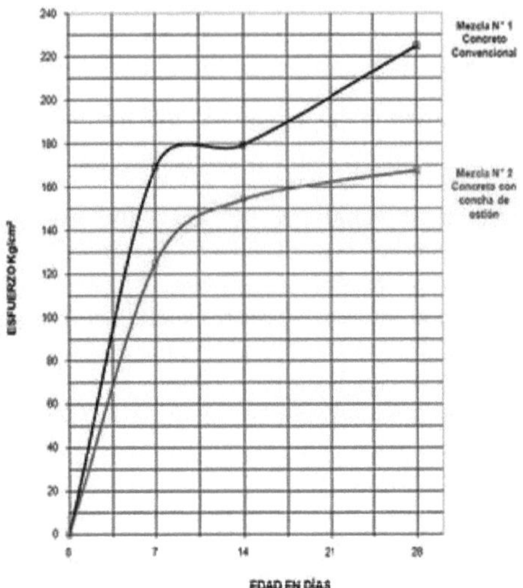

Graph N° 2. Bending stress vs. age in days

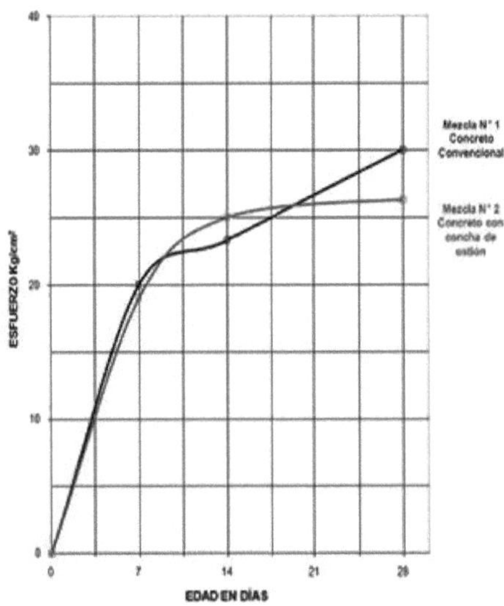

General summary

The first stage of the research consisted of the individual analysis of the fine and coarse aggregates and the recycled material (oyster shell) to obtain their characteristics and to design the concrete mix. In order to obtain representative specimens of the concrete mix, 12 cylindrical specimens of *0= 6"* h=12" and 9 specimens sampled in rectangular prismatic moulds of 15 x 15 x 60 cm were sampled according to the NMX-C-161- 1997-ONNCCE standard; the samples were identified and protected from evaporation. The specimens were cured in the laboratory by total immersion in water, the test ages established were 7, 14 and 28 days.

They were sulphur-headed according to the NMXC-109- ONNCCE standard, verifying the flatness and perpendicularity tolerance parameters. The volumetric masses obtained from samples 1 and 2 were 2296 and 2290 kg/m3 , respectively. The compressive stresses at the age of 28 days for specimen No. 1 corresponding to the concrete mix without recycled material are in the order of 225 kg/cm2 , while the flexural stresses were 30.00 kg/cm2 . In the case of sample No. 2, the maximum compressive stress obtained at the age of 28 days is 170 kg/cm2 and 26.67 kg/cm2 in bending.

Conclusions

A conventional concrete under the design described in this document achieved compressive stresses in the order of 220 kg/cm2 , as well as a bending stress of 30 kg/cm2 . With the addition of oyster shell to a conventional concrete, compression values of 170 kg/cm2 and bending stresses of 27 kg/cm2 were presented. Given the above, we can conclude that the greatest contribution of stresses in the samples with oyster shell is generated in the first 7 days, after which the concrete only acquires 25% of its final strength. The addition of the recycled material reduces the compressive and flexural stresses in the specimens by approximately 23%.

List of references

CONAPESCA. (2015). Specific consultation by production. Available at: http://www.conapesca.

sagarpa.gob.mx/wb/cona/consulta_especifica_por_produccion [15 June 2015].

Essen (2005). "Mycroscopy of historic mortars - a review. Cement and Concrete Research. 1-9.

Government of Spain (n.d.). Crassostrea angulata (Lamarck, 1819).

Available at: http://www. ictioterm.es/scientific_name.php?nc=202

Lovatelli, Farias, and Uriarte (2007). FAO Fisheries and Aquaculture Proceedings, ISSN 2071-1026. Current Status of Bivalve Mollusc Culture and Management and its Future Projection.

Marín, Luquet, Marie, and Medacovic (2008). "Molluscan Shell Proteins: Primary Structure, Origin, and Evolution". *UMR CNRS 5561 'Bioge'osciences,' Universite'de Bourgogne Boulevard Gabriel, 21000 Dijon, France Centerfor Marine Research Rovinj, Ruder Boskovic Institute Giordano Paliaga, 52210 Rovinj, Croatia.

Nguyen, Boutouil, Sebaibi, Leleyter, and Baraud (2013). "Valorization of seashell by-products in pervious concrete pavers", Construction and Building Materials 49, 2013, pp. 151-160. 20_Current environmental research II.indd 184 16/02/21 12:37 185

Draft Mexican Standard PROY-NMX-C-021-ONNCCE-2001 (Construction industry - Cement for masonry (mortar) - Specification and test methods.

Reguero, and García-Cubas (1991). Molluscs of the Shrimp Lagoon, Veracruz, Mexico: Systematics and Ecology. Anales del Instituto de Ciencias del Mar y Limnología.

Schankrania (n.d.). Curing of fresh concrete. Available at: http://www.arqhys.com/ hidratacion-concreto.html.

Webography

https://www.espacioimasd.unach.mx/articulos/num12/reuso_de_desechos_d e_shells_of_ostion.php

http://www.personal.us.es/falejan/Propiedades%20de%20los%20morteros.p df3

https://www.aquahoy.com/noticias/moluscos/30248-exploran-mediante- the-nano-technology-benefits-of-the-benefits-of-ostion-shell

http://fondecyt.gob.pe/ciencia-al-dia/peru-usan-restos-de-conchas-de- fan-for-produce-concrete

https://www.onncce.org.mx/es/venta-normas/catalogo-de-norma

Printed by Books on Demand GmbH, Norderstedt / Germany